"REAL" Junior Doctor Survival Guide

"REAL" Junior Doctor Survival Guide

Written by
Dr Ricky Frazer

Series Editor
John Frazer

First Edition 2011

ISBN : 978-1-4477-6910-1

Preface

It is with immense excitement that I present to you the first edition of this book. It is now over a year since I started writing this book. In the conception of this book I envisioned a text that is presented in an easily accessible format containing the information really needed to survive that first month. The drive for the book came simply from my own experience as a junior doctor. Having received many awards as an undergraduate medical student, graduating with honours and having previously obtained a first class honours degree in Molecular and Cellular Biology I was confident that junior doctor life would be straight forward. On day 1 however much of that knowledge was redundant and I found myself trying to tread water with little knowledge on how to manage simple conditions well. It is with this in mind that I have written this book to ease the transition between student to doctor. I hope with all sincerity that you enjoy reading it as much as I did writing it.

This book would not have been possible without the invaluable enthusiasm and unwavering support of the people close to me. Firstly I would like to thank my brother John for all his time and effort in editing and designing this book and my father John who continues to be the backbone for all I do. I would also like to thank my immediate family and friends, Louise, Gareth, Keira, Carol, Jill, Pat, Nessa, Steve, Holly and Helen for not allowing me to forget about the things that are really important in life when I find myself lost in medicine! I would like to give a special mention to John, Abi, Alex and Raj for making my five years at medical school more fun than it should ever have been! I would also like to thank all of the teachers, mentors and students from Barry Comprehensive School, University of Bath, University Hospital of Wales, University Hospital Llandough and Rockingham General Hospital in Western Australia who without them, quite simply, I would not be where I am today. Finally I would like to thank my mother Sharon who despite adversity continues to be my main inspiration every day.

Ricky Frazer
June 2011

Contents

Introduction

So you passed...well done...and you have had a lovely summer's break, your family are so proud.... "Jimmy's a doctor don't you know?"

Except you know that revision, which let's be honest, became a case of who could do the most multiple choice questions before the exam, well life as a doctor isn't quite like that......unfortunately you don't get a best of five answers to choose from when patients present to the admissions unit...and the pathophysiology of TTP actually isn't essential knowledge anymore!

Your first day as a junior doctor is in many ways like a first sexual experience, its hyped up for many years, it often leaves you embarrassed, passes in a blur, was nothing like you had imagined and was way more tiring than you thought it would be!

Oh and you paid the GMC several hundred pounds for the privilege!

It's with that first time in mind that this book hopes to help prepare you and teach you what you really need to survive...for your first day...not for your first sexual experience...in an enjoyable and informative way!
I recognise that guidelines change, but the basic knowledge and principles fortunately do not.

Anyway, foreplay over!

I shall teach you in a step by step fashion, in the privacy of your own home. This book is a stand-alone piece of work, you do not need any other materials (other than a pen to complete it!) and by the time you reach the end you will have all the skills and knowledge needed to truly kick some sick patient ass!

How to use this book

I am pleased to say that this book is not just a random collection of questions and thus should not be treated as such; rather it is a complete and structured course. If you start on page 1 and work slowly but surely through to the end you will have learned not just the knowledge required but a strategy of how to survive. Randomly doing a question from the workbook is essentially like putting a grey cannula in a frail old lady... pointless!

Many hours have been spent in preparing this workbook.
- I have assumed reasonable background knowledge. You do not need any other books to complete this workbook, that said if you hit upon a topic that you remember little about, then refreshing your knowledge before that summers day in August is probably not the worst idea you ever had!
- I have emphasised the diagnosis that commonly presents in those early few days.
- I am aware that your knowledge of the weird and wonderful is probably better than your knowledge of heart failure. With this in mind, short texts on the subjects are provided where your knowledge may be particularly thin!
- Where appropriate, a summary of the important literature is provided. You can be sure that some cocky SHO (like me!) who has been doing it wrong for two years WILL ask you to justify what you did, worse still some registrar may want you to do a case based discussion on it!

The approach used is based on sound educational theory. Different people learn in different ways...FACT! Learning one way is damn right boring. Therefore, pictures, poems, puzzles, tables, equations and historical vignettes are all found in the text. Occasionally there may even be a joke or two, do not skip over these, some reinforce facts you have already learnt, some may just make you smile, both are needed now and in life in general.

So that's it, we are ready to go...and remember everything you learn here will help make that first time more pleasurable, so for the love of medicine...ENJOY!

1
Farmercowlogy

Farmercowlogy

Drugs

> A junior doctor, lets call him Ricky, on his first day is asked to prescribe paracetamol. He is not quite sure of the dose or the frequency, but is safe in the knowledge that nurses ask for it a lot and are quick to tell him that its 1 gram every 4 hours. Ricky, who was always quite proud of his mental arithmetic, knows that 24/4 = 6 and happy with this minor victory prescribes the paracetamol. However later that day that bloody bleep thing goes off (the novelty of having a bleep wears off pretty damn quickly) and it's the very sweet but a little concerned ward pharmacist Helen. She's wondering ever so politely if you in fact graduated from medical school as you have just overdosed a patient on the most simple of drugs. Turns out 4 hourly is the minimum time gap that they need to be taken apart and four times daily (or QDS in cool doctor talk) is actually the maximum frequency. A very embarrassing, but all the same useful point is learnt...and how I wish that I could say I made that story up....I CANT!

Here follows a list that simply you just need to know...once you think you have learnt them all and you can write them out without any mistakes, write them up in the box provided. All of the drugs on the next page (except a phosphate enema!) can be given orally (PO). For now that's enough to know about, as time goes by you will start to learn ones that can be given by other routes.

Using the bubbles below complete the table so that the right abbreviation matches with the right frequency.

Once Daily	
Once Morning	
Once Nightly	
Twice Daily	
Three Times Daily	
Four Times Daily	
As Required	

OM   PRN   BD   QDS   OD   ON   TDS

Once Daily	OD
Once Morning	OM
Once Nightly	ON
Twice Daily	BD
Three times Daily	TDS
Four times Daily	QDS
As required	PRN

Common Drugs:

Pain Relief

- ✓ Paracetamol 1g QDS

- ✓ Codeine 60mg QDS

- ✓ Co-codamol 8/500 or 30/500 2 tabs QDS (it sounds obvious I know but you can't have co-codamol and paracetamol/codeine on the same chart it's an overdose!!! Who would be stupid enough to do this I hear you cry...trust me!! By the way this error could make a good audit for that CV of yours!

- ✓ Diclofenac 50mg TDS (good for renal colic – not so nice for the stomach and kidneys in general – be kind)

Constipation – may be related to some of those drugs you prescribed above

- ✓ Senna tabs 2 BD (avoid writing senna and magnesium hydroxide 10ml + 10ml like I do and most other doctors) because the pharmacist will whinge about the price...trust me!

- ✓ Lactulose 30ml TDS (should not be used routinely but excellent in alcoholic liver disease, titrated to produce 2/3 stools per day you can even write this in the instruction box to impress the pharmacist)

- ✓ Phosphate enema T PRN (Usually once daily as it often does the trick if you know what I mean)

Nausea/Vomiting

- ✓ Metoclopramide 10mg TDS (incidentally good after too much kebab and alcohol – helps move things through. Don't forget avoid in young women – oculogyric crisis alert!)

- ✓ Cyclizine 50mg TDS

Night Sedation

- ✓ Temazepam 10mg (Give 5mg in elderly) – don't be bullied into giving this – unless of course the nurse looks scary...

Some other common ones you may already know! Have a go – answers below.

Drugs	Dose	Frequency
Furosemide		
Simvastatin		
Aspirin		
Dipyridamole		
Co-amoxiclav		
Ciprofloxacin		

Of course this list is not exhaustive, but it will save you hours of time if you arrive on day one armed with these doses.

Drugs	Dose	Frequency
Furosemide	40mg	OD
Simvastatin	40mg	ON
Aspirin	75mg	OD
Dipyridamole	200mg	BD
Co-amoxiclav	625mg	TDS
Ciprofloxacin	500mg	BD

Here's a quick joke...

"Yeah, Doc, what's the news?" answered Fred when his doctor called with his test results. "I have some bad news and some really bad news," admitted the doctor. "The bad news is that you only have twenty-four hours to live."
"Oh my god," gasped Fred, sinking to his knees. "What could be worse news than that?"
"I couldn't get hold of you yesterday."

2

Wheezy Peezy

Wheezy Peezy

Asthma

I want you to close your eyes for a moment at the end of this first paragraph...imagine a day gone by that you have spent in the medical admissions unit....its frantic down there, you are the house officer, you have been asked to check blood results, see another patient and the young girl with asthma is getting worse, she is finding it difficult to speak, bradycardic, and you appear to be on your own. It's too much for you to cope. Close your eyes and let the anxious feeling rush through you for a minute or two and feel the butterflies in your belly.

---------------------------Eyes Closed 5 minutes----------------------------

That day will happen, we have all been there, but relax, when you experience that feeling again, the difference is you will be armed with the information below, ready to do the simple things well and ready to save that girls life!

Let's go....

Q.1

A 21 year old female of Japanese origin lives with her grand parents. She has a 4 year old brother who has recently suffered an acute coryzal illness. Her mother is a nursery school teacher whilst her dad is a successful lawyer. The family returned from Japan only 3 months ago having spent a week in an orphanage in Tokyo. The girl presents to MAU feeling more SOB for 5 days. 27 days before, her GP prescribed a 1 week course of antibiotics for a throat infection. Her dog also died 4 weeks ago with some unknown infection. She is known to have asthma, she has had a cough for 3 days which is unproductive, and she has felt a little feverish, which she attributes to the fan no longer working in her bedroom. She has had no haemoptysis but does say she has been wheezy. She also tells you that "I feel hungry all the time but I don't know why".

On examination, temp 37.3, heart rate 90, BP 105/75, JVP not elevated, a few polyphonic wheezes in the chest, abdomen soft – no organomegaly, clubbing both hands, PEFR – 85% predicted, heart sounds normal.

Firstly, good effort on your detailed history, the only problem is you now need to work out what the hell is going on... can we blame all her symptoms on her dead dog?! If this was a multiple choice question you would see some of this as meaningless bulk. So what do we know about this patient other than a very detailed family and pet history? What are the most important features of the history that are likely to help with an immediate diagnosis?

List Below
1) _____
2) _____
3) _____
4) _____
5) _____

A.1

1) Known asthma

2) SOB

3) Wheeze

4) Fever

5) Unproductive cough

During your first week on call you are almost certain to admit a young asthmatic patient and this can often be scary! The main problem is these patients can go off very very quickly and the harsh reality is that if your 95 year old patient dies of pneumonia there will be a lot less questions asked than if a 21 year old ends up intubated in ICU. With that in mind... arrrgh... fear not as treatment is in fact very straight forward.

Now using the examination findings, which features indicate the severity of the asthma exacerbation and therefore should guide treatment?

1)
2)
3)
4)

1) PEFR
2) Speech (I know, not always an examination finding!)
3) Heart Rate
4) Respiratory Rate

Now using those criteria see if you can remember how we grade severity of asthma...once you have learnt this, treatment is a doddle!

	Mild – moderate	Acute severe	Life threatening
PEFR			
Speech			
Heart rate			
Respiratory rate			

	Mild -moderate	Acute severe	Life threatening
PEFR	> 50% of predicted	33-50% of predicted	<33% of predicted
Speech	Normal	Unable to complete sentences	Exhaustion or coma
Heart rate	< 110 beats per minute	> 110 beats per minute	bradycardia
Respiratory rate	<25 Breaths/min	> 25 breaths/min	Cyanosis/Silent chest

The golden rule here is that imagine you are the consultant on the post take ward round who has seen many many acute exacerbations of asthma...well those same consultants have used the same books and literature to define the severity and therefore are looking for those key words when you present the case....it's the only way they know what to do or indeed what you will tell them you did for brownie points!

Treatment guidelines – follow these and you really will look awesome!

Make sure you can write these out point by point before you continue to the next section, asthma attacks are one of the few areas where as a junior doctor you can decide if someone lives or goes up to the spirit in the sky!

Acute asthma

- ✓ Start 40 – 60% oxygen
- ✓ Salbutamol 5mg (Nebulised on oxygen)
- ✓ Oral prednisolone 40mg daily for 5 days
- ✓ **Monitor response at 15-30 minutes. If no improvement, treat as life threatening**
- ✓ If improvement only slight, then repeat nebulised salbutamol/ ipratropium bromide
- ✓ If pulse and respiratory rate settling and PEFR > 50% predicted, continue 4 hourly nebulised salbutamol, daily oral prednisolone, step up usual inhaled steroid therapy. If no improvement, treat as life threatening

Life-threatening asthma

- ✓ Inform your senior immediately – ideally the registrar because ICU need to know about this patient
- ✓ Start 40 – 60% oxygen
- ✓ Salbutamol 5mg with ipratropium 500 micrograms (nebulised on 8l/min of oxygen
- ✓ Oral prednisolone 40mg or IV hydrocortisone 100mg stat (or both if patient is very ill), oral steroid is as effective
- ✓ **If no improvement after 15-30 minutes** give nebulised beta-agonist more frequently e.g. salbutamol 5mg up to every 15 minutes – make sure your senior is reviewing Continue nebulised ipatropium 500 micrograms 4-6 hourly
- ✓ **Patient may require IV magnesium sulphate, IV aminophyline or IV B2 agonist – discuss this with your senior! If unwell enough to receive this they need experienced input now!**

Take a break

On your return, you'll need to check you've remembered the information above. Keep practicing until you get it all, you'll need it!

Acute Asthma Treatment

Life threatening Asthma Treatment

--18---

3

Nervous!

Nervous

Transient Ischaemic Attack

You are very unlikely to get through your first on call for medicine without a GP sending in a.............? **TIA**

With the greatest respect to GP'S (who incidentally are underworked, overpaid, and oversexed – ok so I'm not certain about the last one) will send pretty much anything into hospital with this diagnosis, from a slightly confused old lady to a young guy with cramp! In their defence it could be a TIA – but probably isn't. Either way they did their job, it's now up to you to manage them.

Let's consider a common scenario:

Q.2
A 76 year old lady Miss Jones who was previously fit and well presents with a 3 hour history of weakness in the right arm. It started this morning suddenly, the patient was holding a cup of tea and noticed that the arm became weak and she dropped it. On arriving in MAU her weakness has completely resolved and the patient feels stupid that her GP even made her come to the hospital. The patient is in sinus rhythm and cardiovascular and neurological examinations are entirely normal.

Why is this likely to represent a TIA rather than a stroke?

How would you manage this patient?

A.2

1) The first answer here utilises knowledge you are likely to have acquired very early in your undergraduate lectures. 24 – not the series on telly which I must get round to watching, but that by definition the symptoms in a TIA last less than 24 hours and stroke lasts more than 24 hours – simple as that!

2) As for the second answer, well luckily for you our kind friends at NICE have recently updated the guidelines relating to management of TIA's and stroke and as I'm sure I have mentioned before and almost certain to mention again, you can't go far wrong if you treat patients in the way NICE recommend. Furthermore you will be amazed how often this isn't done. Is this because the consultant knows best? Unlikely..... is it because they haven't kept abreast of the recent guidelines? I will leave you to decide.....Anyway where was I...oh yes the guidelines:

NICE Guidelines

They advocated the use of the ABCD2 prognostic score for risk stratifying patients who've had a suspected TIA:

	Criteria	Points
A	Age = 60 years	1
B	Blood pressure = 140/90 mmHg	1
C	Clinical features	
	- Unilateral weakness	2
	- Speech disturbance, no weakness	1
D	Duration of symptoms	
	- > 60 minutes	2
	- 10-59 minutes	1
	Patient has diabetes	1

This gives a total score ranging from 0 to 7. People who have had a suspected TIA who are at a higher risk of stroke (that is, with an ABCD2 score of 4 or above) should receive:

- Aspirin (300 mg daily) started immediately
- Specialist assessment and investigation within 24 hours of onset of symptoms
- Measures for secondary prevention introduced as soon as the diagnosis is confirmed, including discussion of individual risk factors

How did you treat our lady? More importantly how often have you seen these patients be discharged with f/u in a clinic in 4 weeks!

If the ABCD2 risk score is 3 or below:

- Specialist assessment within 1 week of symptom onset, including decision

On brain imaging

- If vascular territory or pathology is uncertain, refer for brain imaging

People with crescendo TIA's (two or more episodes in a week) should be treated as being at high risk of a stroke, even though they may have an ABCD2 score of 3 or below.

In December 2010 NICE published guidance on the use of clopidogrel and modified-release dipyridamole for the prevention of occlusive vascular events (ischaemic stroke, TIA or heart attack). The guidance which updates the previous NICE guidelines published in 2005 recommends:

- Aspirin 300mg for 14 days after TIA or ischaemic stroke – don't forget can be given PR if NBM.

After the 14 days NICE then recommend:

- Clopidogrel 75mg daily for patients who had an ischaemic stroke

- Modified- release dipyridamole 200mg BD plus aspirin 75mg OD as an option for patients who have had a TIA. For patients who had an ischaemic stroke, modified-release dipyridamole plus aspirin should only be used where clopidogrel is contraindicated or not tolerated

- Modified-release dipyridamole 200mg BD alone is an option for patients who had an ischaemic stroke or a TIA, only where treatment with aspirin and clopidogrel is contraindicated or not tolerated

Consider our lady then:

Miss Jones was seen in the rapid access TIA clinic the next day after a clearly wise beyond their years house officer had her seen by the specialist within 24 hours. On further questioning, for the past three weeks the patient describes having episodes of transient loss of vision in the right eye. Carotid ultrasound reveals a 48% stenosis of her left carotid artery and ECG confirms sinus rhythm.

What is the most appropriate anticoagulant management for this patient long term?

Is any further carotid intervention required at this point?

Answers:

1. Clearly this lady has crescendo TIA and therefore by definition is considered to be at high risk. In view of this the patient should be started on long term 75mg of aspirin and modified release 200mg dipyridamole BD

2. With regards to carotid endarterectomy: recommend if patient has suffered stroke or TIA in the carotid territory and are not severely disabled and should only be considered if carotid stenosis > 70%

Time for another joke!

I was sitting in the waiting room of the hospital after my wife had gone into labour and the nurse walked out and said to the man sitting next to me, "Congratulations sir, you're the new father of twins!"

The man replied, "How about that, I work for the Doublemint Chewing Gum Company." The man then followed the woman to his wife's room.

About an hour later, the same nurse entered the waiting room and announced that Mr. Smith's wife has just had triplets. Mr. Smith stood up and said, "Well, how do ya like that, I work for the 3M Company."

The gentleman that was sitting next to me then got up and started to leave. When I asked him why he was leaving, he remarked, "I think I need a breath of fresh air."
The man continued, "I work for 7-UP."

Stroke

Q.3
A 53 year old gentleman is admitted to the emergency department with a right hemiplegia. His symptoms started around 4 hours ago and he has no headache, visual disturbance or loss of consciousness. On examination a dense right hemiplegia is found. Blood pressure is 120/78mmHg. GCS is 15/15 and pupils are equal and reactive to light. An urgent CT scan is performed shortly after his arrival and no abnormality is found.

Does this represent a TIA or a stroke and why?

What is the appropriate initial management in this patient?

What is meant by secondary prevention?

A.3 1) This is a slightly tricky question, but it is not a TIA as yet because as YOU KNOW, symptoms have to have resolved within 24 hours and these have not yet resolved. So actually it could be either a TIA or a stroke.

2) Patient is outside the window for thrombolysis and therefore should be treated with aspirin.

3) Secondary prevention refers to treatment of the patients risk factors after a primary event to help prevent further episodes and complications.

Stroke Management

The Royal College of Physicians (RCP) published guidelines on the diagnosis and management of patients following a stroke. NICE have also issued guidelines.

Selected points relating to the management of acute stroke include:

- Oxygen saturation, blood glucose, hydration and temperature should be maintained within normal limits
- Blood Pressure should not be lowered acutely unless complications develop such as hypertensive encephalopathy
- Once haemorrhagic stroke has been excluded, aspirin 300mg orally or rectally should be given as soon as possible
- If the cholesterol is > 3.5 mmol/l patients should be commenced on a statin
- With regards to atrial fibrillation, the RCP state 'anticoagulants should not be started until brain imaging has excluded haemorrhage and usually not until 14 days have passed from onset of the stroke'

You will hopefully see with increased frequency that thrombolysis is being prescribed in the acute setting for these patients.

The current guidelines for patients receiving thrombolysis are:

1) It is administered within 3 hours of onset of symptoms suggestive of stroke (unless part of a clinical trial)
2) Haemorrhage has been excluded (i.e. imaging has been performed)

Time for a little WORDSEARCH over a tea/coffee/cheeky alcoholic beverage.

Find the risk factors that you would want to address in a patient presenting with either a TIA or stroke. They can occur backwards, forwards, diagonally, horizontally or vertically.

H	L	O	B	S	L	D	E	K	G	E	H
E	I	E	I	O	E	G	T	L	O	Y	A
A	E	I	E	S	N	A	W	L	P	T	A
P	E	O	B	I	E	H	R	E	K	P	L
O	E	O	K	D	K	T	R	I	T	U	C
C	H	O	L	E	S	T	E	R	O	L	O
O	M	I	T	S	E	I	E	B	G	O	H
S	H	H	W	N	D	N	O	T	A	N	O
T	O	T	S	I	R	S	D	Y	T	I	L
E	O	I	M	R	N	A	A	L	H	L	D
H	O	A	T	B	E	R	P	A	R	H	R
N	E	Y	T	S	T	E	R	A	O	A	W

Take a break!

4

Metabollocks and Endocrap

Hyperkalaemia

A very common problem you will face in your day to day jobs is knowing what to do with abnormal blood results – assuming of course you can actually get a phlebotomist to take them. I recall a Christmas weekend where a little FY2 was left to take 27 bloods each day because the phlebotomist took 5 days off with no plan for the 1200 patients they left behind!

Anyhow...

A very common abnormal blood result you will encounter is hyperkalaemia.

Hyperkalaemia – is a source of much concern amongst nurses and doctors and this is because untreated it can cause life threatening arrhythmias. However as I'm sure you remember the differential for hyperkalaemia is by no means small and a common cause is a haemolysed sample. Therefore whilst I am not suggesting that you shouldn't treat it I do recommend that you at least send off a repeat sample – if only because you can give your colleagues the speech "the amount of times you just repeat the sample and it comes back normal" – I love that speech!

List the causes of Hyperkalaemia

○ Acute renal failure
○ Drugs - Potassium sparing diuretics, ACE inhibitors, angiotensin 2 receptor blockers, spironolactone, ciclosporin
○ Metabolic acidosis
○ Addison's
○ Rhabdomyolysis
○ Massive blood transfusion
○ And my favourite - haemolysis

As you can see from the answers above, potassium levels are regulated by a number of factors including acid base balance, aldosterone and insulin levels. For your knowledge, the reason metabolic acidosis is associated with hyperkalaemia is that hydrogen and potassium ions compete with each other for exchange with sodium ions across the distal tubule.

Management

Whilst you're waiting for the sample to return, management may be categorised by the aims of the treatment.

- Stabilisation of the cardiac membrane
  - Intravenous calcium gluconate

- Short term shift in potassium from extracellular to intracellular fluid compartment
  - Combined insulin/dextrose infusion
  - Nebulised salbutamol

- Removal of potassium from the body
  - Calcium resonium (orally or enema – oral nicer!)
  - Loop diuretics
  - Dialysis

It really is as simple as ABC

A – Astabilisation approach
B – Bang it into the cells
C – Clear it out of the body

(OK, admittedly not the best aid memoire that I have ever provided!)
Consider the ECG below

Can you identify the characteristic changes seen in hyperkalaemia?

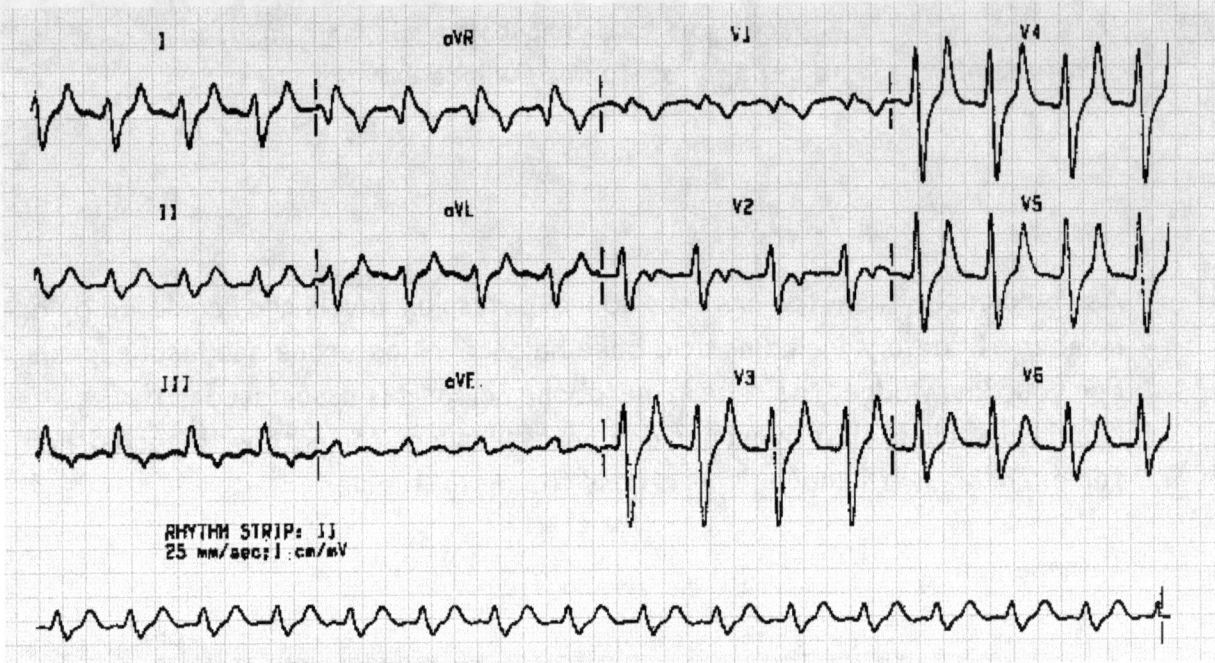

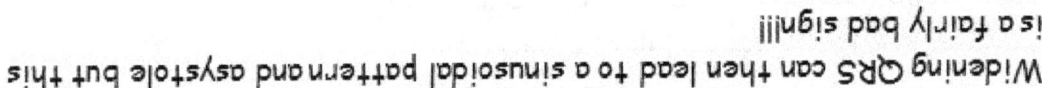

Answer

Tall T waves, small p waves, widening QRS

Widening QRS can then lead to a sinusoidal pattern and asystole but this is a fairly bad sign!!!

Crossword (for a bit of fun!)

See how many fruits or vegetables you can name that can increase your potassium

ACROSS
1 ... and Lemons sing the bells of St. Clements!
4 The name comes from the Nahuatl word ahuacatl ('testicle', a reference to the shape of the fruit)!
5 Also known as the Chinese gooseberry!
6 Don't slip on the skins!

DOWN
2 Popeye!
3 Are they a fruit or a vegetable?

Solution:

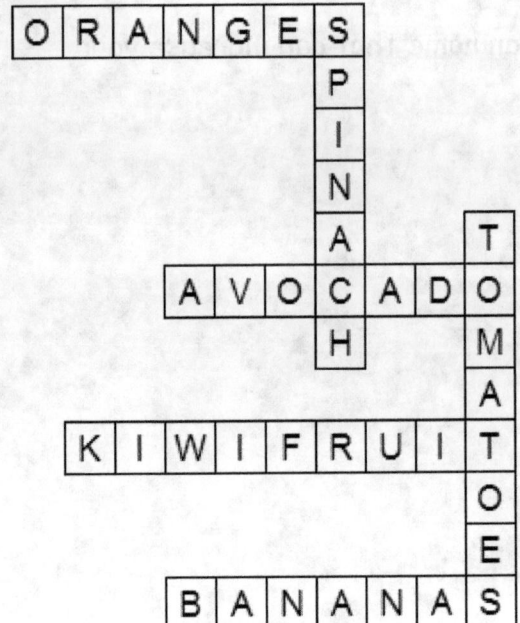

Diabetic Ketoacidosis (DKA)

DKA is almost the holy grail of junior doctor management. These patients are SICK (and not sick as in awesome – wow I'm down with the kids) no, sick as in, need to help them now sick!

Two years ago and sadly still today it's not uncommon for patients to be managed with ABG monitoring and a sliding scale! However in the last year DKA management has been overhauled and a lot my dear friend has changed.

Therefore I prepare you for possibly the trickiest and longest section of this book so far... that said, knowing this on your first day will quite simply leave you head and shoulders above everyone else.

So here goes...

First of all... reminder of some basic pathophysiology.

DKA is a complex disordered metabolic state characterised by hyperglycaemia, acidosis, and ketonaemia. DKA usually occurs as a consequence of absolute or relative insulin deficiency that is accompanied by an increase in counterregulatory hormones (i.e., glucagon, cortisol, growth hormone, epinephrine). This type of hormonal imbalance enhances hepatic gluconeogenesis and glycogenolysis resulting in severe hyperglycaemia. Enhanced lipolysis increases serum free fatty acids that are then metabolised as an alternative energy source in the process of ketogenesis. This results in accumulation of large quantities of ketone bodies and subsequent metabolic acidosis. Ketones include acetone, 3-beta-hydroxybutyrate, and acetoacetate. The predominant ketone is 3-beta-hydroxybutyrate.

Q.4

A 22 year old female with a history of type 1 diabetes mellitus, presents to the emergency department with abdominal pain and vomiting. Finger prick testing estimates the blood glucose to be 25mmol/l. You perform an ABG and find the pH to be 7.21 and the bicarbonate to be 8mmol/l. On examination the patient is clinically dehydrated and weighs 82kg. You insert a nice green venflon and take some bloods. 1 litre of 0.9% saline is infused and a sliding scale is commenced.

What rate should the insulin initially be given?

What are the diagnostic criteria for DKA?

What parameters are indicative of severe DKA and necessitate immediate senior review and possible ICU admission?

What is the rate of initial fluid replacement?

Changes in management

Until recently, management of DKA has focussed on lowering the elevated
blood glucose with fluids and insulin, using arterial pH and serum bicarbonate
to assess metabolic improvement.
Recent developments now allow us to focus on the underlying metabolic
abnormality (ketonaemia).

• Measurement of blood ketones, venous (not arterial) pH and bicarbonate
and their use as treatment markers

• Monitoring of ketones and glucose using bedside meters

• Replacing 'sliding scale' insulin with weight-based fixed rate intravenous
insulin infusion (IVII)

• Use of venous blood rather than arterial blood in blood gas analysers

• Monitoring of electrolytes on the blood gas analyser with intermittent
laboratory confirmation

• Continuation of long acting insulin analogues (Lantus® or Levemir®) as
normal

Management pathway

After your initial assessment of the patient (see answer to question 3), it's
time to treat! The following algorithm has been adapted from the
diabetes.org.uk website which is defiantly worth a look! Here we will focus on
acute management...the first 6 hours...however clearly patient management
does not end until the heroic house officer has saved the day!

Initial management - The first hour - Your time to shine

Action 1: Commence 0.9% sodium chloride solution
(use large bore cannula) via infusion pump

See answer to question four for rate of fluid replacement.

Action 2: Commence a fixed rate intravenous insulin infusion (IVII)
(0.1unit/kg/hr based on estimate of weight) 50 units human soluble insulin
(Actrapid® or Humulin S®) made up to 50ml with 0.9% sodium chloride
solution. If patient normally takes long acting insulin analogue (Lantus®,
Levemir®) continue at usual dose and time

Action 3: Assess patient
· Respiratory rate; temperature; blood pressure; pulse; oxygen
saturation · Glasgow Coma Scale · Full clinical examination

Action 4: Further investigations
· Capillary and laboratory glucose · VBG · U & E · FBC · Blood cultures
· ECG · CXR · MSU

Action 5: Establish monitoring regimen
· Hourly capillary blood glucose · Hourly capillary ketone measurement if
available · Venous bicarbonate and potassium at 60 minutes, 2 hours and 2
hourly thereafter · 4 hourly plasma electrolytes · Continuous cardiac
monitoring if required · Continuous pulse oximetry if required

Action 6: Consider any precipitating causes and treat appropriately

60 Minutes - 6 Hours - Almost a hero!

Aims of treatment:

· Rate of fall of ketones of at least 0.5 mmol/L/hr OR bicarbonate rise 3
mmol/L/hr and blood glucose fall 3 mmol/L/hr
· Maintain serum potassium in normal range · Avoid hypoglycaemia

Action 1: Re-assess patient, monitor vital signs
· Hourly blood glucose (lab blood glucose if meter reading 'HI')
· Hourly blood ketones if meter available

• Venous blood gas for pH, bicarbonate and potassium at 60 minutes, 2 hours and 2 hourly thereafter
• If potassium is outside normal range, re-assess potassium replacement and check hourly. If abnormal after further hour seek immediate senior medical advice

Action 2: Continue fluid replacement via infusion pump as follows:
• 0.9% sodium chloride 1L with potassium chloride over next 2 hours
• 0.9% sodium chloride 1L with potassium chloride over next 2 hours
• 0.9% sodium chloride 1L with potassium chloride over next 4 hours
• Add 10% glucose 125ml/hr if blood glucose falls below 14 mmol/L

More cautious fluid replacement in young people aged 18-25 years, elderly, pregnant, heart or renal failure. (Consider HDU and/or central line).

Action 3: Assess response to treatment
Insulin infusion rate may need review if:
• Capillary ketones not falling by at least 0.5 mmol/L/hr • Venous bicarbonate not rising by at least 3 mmol/L/hr • Plasma glucose not falling by at least 3 mmol/L/hr • Continue fixed rate IVII until ketones less than 0.3 mmol/L, venous pH over 7.3 and/or venous bicarbonate over 18 mmol/L

If ketones and glucose are not falling as expected always check the insulin infusion pump is working and connected and that the correct insulin residual volume is present (to check for pump malfunction).

If equipment working but response to treatment inadequate, increase insulin infusion rate by 1 unit/hr increments hourly until targets achieved.

A.4

1) 0.1mg/kg/hour

2) Capillary blood glucose above 11mmol/L
   Capillary ketones above 3mmol/l or urine ketones 2+ or more
   Venous pH less than 7.3 and/or bicarbonate less than 15mmol/l

3) Presence of one or more:
   Blood ketones over 6mmol/L, Bicarbonate below 5mmol/L, Venous pH below 7.1
   Hypokalaemia on admission (less than 3.5mmol/L)
   GCS less than 12
   Oxygen sats less than 92%
   Systolic BP below 90mmHg

4) Systolic BP (SBP) below 90mmHg

   Give 500ml of 0.9% sodium chloride solution over 10-15 minutes.
   If SBP remains below 90mmHg repeat whilst requesting senior
   input. Most patients require between 500 to 1000ml given
   rapidly.
   Consider involving the ITU/critical care team.
   Once SBP above 90mmHg give 1000ml 0.9% sodium chloride over
   next 60 minutes. Addition of potassium likely to be required in
   this second litre of fluid.

   Systolic BP on admission 90 mmHg and over give 1000ml 0.9%
   sodium chloride over first 60 minutes.

   Potassium replacement:
   > 5.5 - Nil
   3.5-5.5 - 40 mmol/L
   < 3.5 - senior review – additional potassium required

Additional measures:

• Regular observations and Early Warning Score (EWS)
• Accurate fluid balance chart, minimum urine output 0.5ml/kg/hr
• Consider urinary catheterisation if incontinent or anuric (not passed urine
by 60 minutes) • Nasogastric tube with airway protection if patient obtunded
or persistently vomiting
• Measure arterial blood gases and repeat chest radiograph if oxygen
saturation less than 92% • Thromboprophylaxis with low molecular weight
heparin
• Consider ECG monitoring if potassium abnormal or concerns about cardiac
status

As so often is the case in medicine you are likely to have a senior tell you that you still need to do an ABG "because venous PH and bicarbonate aren't accurate" Ready for this very moment, below is your polite justification as to why you're managing this condition correctly.

"The difference between venous and arterial pH is 0.02-0.15 pH units and the difference between arterial and venous bicarbonate is 1.88 mmol/L. This will change neither diagnosis nor management of DKA".

Right that's it... DKA in a nutshell!! Firstly WELL DONE on getting through that section....you won't possibly remember all of the information first time but for now take away the main principles and never be scared to ask a senior for help or for an explanation if you find the practice at your hospital does not represent the guidelines above.

Great news... Time for a joke!

Hank Smith gets home from work one day and finds his wife has been crying. "What's wrong?" he asks."Hank, promise you won't get mad, but I went to see the new doctor today and he told me I've got a pretty pussy."
"WHAT?" he shouts. With that he grabs a baseball bat from the cupboard and storms down to the doctor's office and through the reception area.
Without knocking he bursts into the doctor's office. The doctor is in the process of giving an old lady a breast examination. She screams and tries to cover herself.
Without waiting, Mr. Smith charges up to the doctor, smashes the baseball bat down on the desk and says, "You flaming pervert! How dare you say my wife has a pretty pussy!" The doctor replies, "I'm sorry Mr. Smith, but there has been a misunderstanding. I only told your wife that she has Acute Angina."

5

Shits and Giggles

Gastroenterology

Diarrhoea

There are no two ways about it, hospitals are smelly kinds of places and diarrhoea is a way of life.

Two things you must be able to do before you get to the ward:

1) Spell diarrhoea
2) Have a differential for diarrhoea

Ok, back to detention at school – you've been naughty and asked to write lines. You have 20 seconds to write out diarrhoea correctly as many times as you can – time yourself... my best try is 9.

Ok, I know that was a silly way to spend 20 seconds or possibly a lot longer if you tried to better mine!

Second task is a differential for diarrhoea. In order to make this memorable see if you can use the word DIARRHOEA as a mnemonic.

Mnemonic

D
I
A
R
R
H
O
E
A

Here is one I made earlier if you couldn't make your own.

Upside down Diet, Infective, Autoimmune, Retention (leading to overflow), Radiation, Hormonal (remember those examination question on VIPOMA), Other GI problems (Crohn's, UC) Endocrine and Antibiotics.

Unfortunately the last of these is becoming an increasing problem due to infection with Clostridium difficile.

Have a go at the next problem and delete the appropriate words from the sentence below.

Clostridium difficile is a gram positive/negative rod often encountered in the community/hospital environment. It produces an exotoxin/endotoxin which causes intestinal damage leading to a syndrome called pseudomembranous colitis. Clostridium difficile develops when the normal gut flora are suppressed by broad/narrow spectrum antibiotics. Clindamycin/erythromycin was historically associated with causing Clostridium difficile but the aetiology has evolved significantly over the past 10 years. Second and third generation cephalosporin's are now the leading causes.

NO CHEATING!!

Here is how it should read now:

Clostridium difficile is a gram positive rod often encountered in the hospital environment. It produces an exotoxin which causes intestinal damage leading to a syndrome called pseudomembranous colitis. Clostridium difficile develops when the normal gut flora are suppressed by broad spectrum antibiotics. Clindamycin was historically associated with causing Clostridium difficile but the aetiology has evolved significantly over the past 10 years. Second and third generation cephalosporin's are now the leading causes.

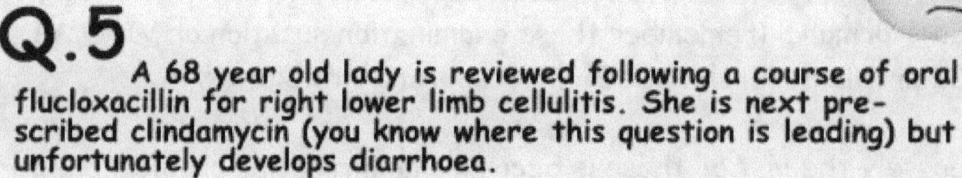

Q.5

A 68 year old lady is reviewed following a course of oral flucloxacillin for right lower limb cellulitis. She is next prescribed clindamycin (you know where this question is leading) but unfortunately develops diarrhoea.

Name three features of Clostridium difficile other than diarrhoea?

What test is used to confirm the diagnosis of Clostridium difficile?

Which antibiotics are used in the management of Clostridium difficile?

A.5

1)   Features - Diarrhoea, Abdominal Pain, Raised White Cell Count, Toxic Megacolon if infection severe

2)   Diagnosis - Detection of Clostridium difficile toxin in the stool

3)   Management - First line therapy is oral metronidazole for 10 – 14 days, if severe or not responding to metronidazole then oral vancomycin may be used. For life threatening infections a combination of oral vancomycin and intravenous metronidazole should be used

6

Slice and Dice

Slice and Dice

Renal Stones

A little starter for ten, a multiple choice question.

To prevent calcium renal stones which one of the following may be useful?

A) Pyridoxine

B) Allopurinol

C) Thiazide Diuretics

D) Ferrous Sulphate

E) Lithium

This is actually a nice question, the fact that thiazide diuretics cause hypercalcaemia is sometimes confused with their role in preventing calcium renal stones – the hypercalcaemia seen is secondary to increased distal tubular calcium resorption and hence lower calcium concentration in the urine.

A common surgical presenting complaint is renal stones and the most common cause is as mentioned above, calcium stones. I include this because there is a magic bullet that all junior doctors should know about and that is diclofenac 75mg.

Renal colic is one of those rare times that patients will not object to PR medication, renal colic hurts really really bad! Diclofenac is like a magic trick, much like the relief of having a catheter inserted when in acute urinary retention. Aw bliss!

Acute Management

- Diclofenac 75mg
- Stones < 5mm will usually pass spontaneously
- Lithotripsy, nephrolithotomy may be required

Prevention of renal stones

Calcium Stones
- Thiazide diuretics (increased distal tubular calcium resorption)
- High fluid intake

Oxalate stones
- Cholestyramine reduces urinary oxalate secretion
- Pyridoxine reduces urinary oxalate secretion

Uric acid stones
- Allopurinol
- Urinary Alkalinisation e.g. Oral Bicarbonate

Time for a cuppa with a biscuit and a light-hearted read.

Top ten things you don't want to hear in surgery:

1 Don't worry. I think it is sharp enough

2 Nurse, did this patient sign the organs donation card?

3 Damn! Page 84 of the manual is missing

4 Everybody stand back! I lost a contact lens

5 Hand me that...uh...that uh.....thingie

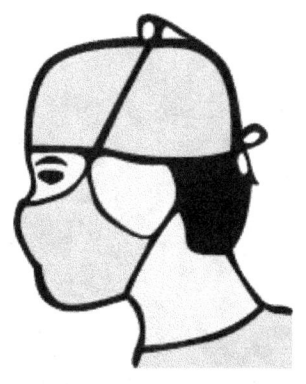

6 Better save that. We'll need it for the autopsy

7 Accept this sacrifice, O Great Lord of Darkness

8 Whoa, wait a minute, if this is his spleen, then what's that?

9 Ya know, there's big money in kidneys. Hell, he's got two of'em

10 What do you mean you want a divorce?

7

Restbite

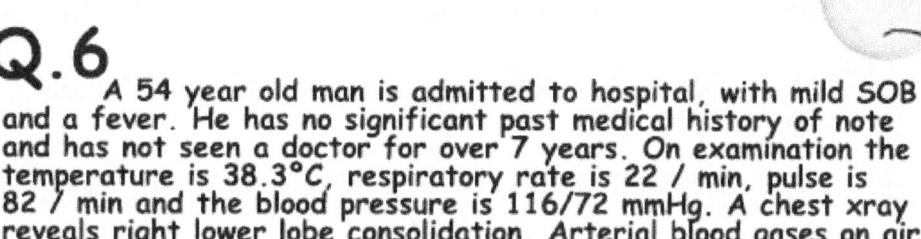

Q.6

A 54 year old man is admitted to hospital, with mild SOB and a fever. He has no significant past medical history of note and has not seen a doctor for over 7 years. On examination the temperature is 38.3°C, respiratory rate is 22 / min, pulse is 82 / min and the blood pressure is 116/72 mmHg. A chest xray reveals right lower lobe consolidation. Arterial blood gases on air are as follows:

pH 7.39
pCO_2 4.6 kPa
pO_2 9.8 kPa

What is the most appropriate antibiotic therapy in this patient?

Which prognostic factors determine severity and how does this influence management?

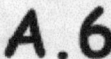

A.6

1) Amoxicillin

2) Prognostic factors:

CURB – 65 criteria of severe pneumonia

- Confusion – abbreviated mental test score < 8 or new disorientation in person place or time
- Urea > 7 mmol/l
- Respiratory Rate > 30 breaths/ min
- Blood Pressure (SBP < 90 or DBP ≤ 60)
- Age ≥ 65

Despite the love of doctors for Co-amoxiclav, the 2004 British Thoracic Society guidelines recommend oral amoxicillin as the first line antibiotic for hospitalised patients with non-severe community acquired pneumonia (CURB 0-1). This applies if the patient has not yet been treated in the community or, as is unfortunately a reality in the medical admissions department, the patient has been admitted for non-clinical reasons.

To determine severity and in turn to guide treatment, it is important in patients presenting with pneumonia to perform a severity assessment... importantly for you it looks awesome on a post take ward round when you summarise with "This patient presented with a severe pneumonia and in keeping with the CURB 65 score of 3 I have treated with..."

On commencing your first house officer job, it is worth spending a few minutes familiarising yourself with your own hospital protocols as these often differ between different trusts based on resistance and sensitivity patterns.

Score 1 point for each feature present

Patients with 3 or more (out of 5) of the above criteria are regarded as having a severe pneumonia.

Management of severe pneumonia

These patients usually require IV antibiotics with Co-amoxiclav 1.2g TDS supplemented with oral clarithromycin 500mg BD for 10 days.

They should also be reviewed by your senior to see if they are a candidate for ICU particularly if hypoxic, hypercapnic or acidotic.

Remember - these patients should have a follow up CXR at 6 weeks to exclude lung cancer - this very rarely happens!!!!

Patients who score 2 should be considered for hospital supervised treatment, perhaps as a short stay patient! You can use your exceptional clinical acumen to make the decision.

Make sure you can write out the CURB65 score from memory before you move on and remember to document it every time you see a patient with pneumonia... you could even audit it! Just a thought!!!!

CURB65

Time for some fun!

Use the anagrams below to write a list of the most common causes of community acquired pneumonia.

1. Seetcpnecus uotpcomoniar =
2. Hopiluainzs fluenhaeem =
3. Socoreccalt auusaphyl =
4. Mcameupna msyoniaeopl =
5. Ksielpnleo euniaeblam =

Answers:
1) Streptococcus pneumoniae
2) Haemophilus influenzae
3) Staphylococcus aureus
4) Mycoplasma pneumoniae
5) Klebsiellla pneumoniae

Mix and match – match the bacterial cause of pneumonia with the association below:

1) Alcoholic 2) Young 3) Herpes labialis 4) COPD 5) IV drug user

Bacteria	Association
Streptococcus pneumoniae	
Haemophilus influenzae	
Staphylococcus aureus	
Mycoplasma pneumoniae	
Klebsiellla pneumoniae	

Joke time!

No book would be complete without some classic "Doctor Doctor" jokes.

Doctor, Doctor my son has swallowed my pen, what should I do?
Use a pencil 'till I get there

Doctor, Doctor I think I'm suffering from Déjà Vu!
Didn't I see you yesterday?

Doctor, Doctor I've got wind! Can you give me something?
Yes - here's a kite!

8

Heart and Soul

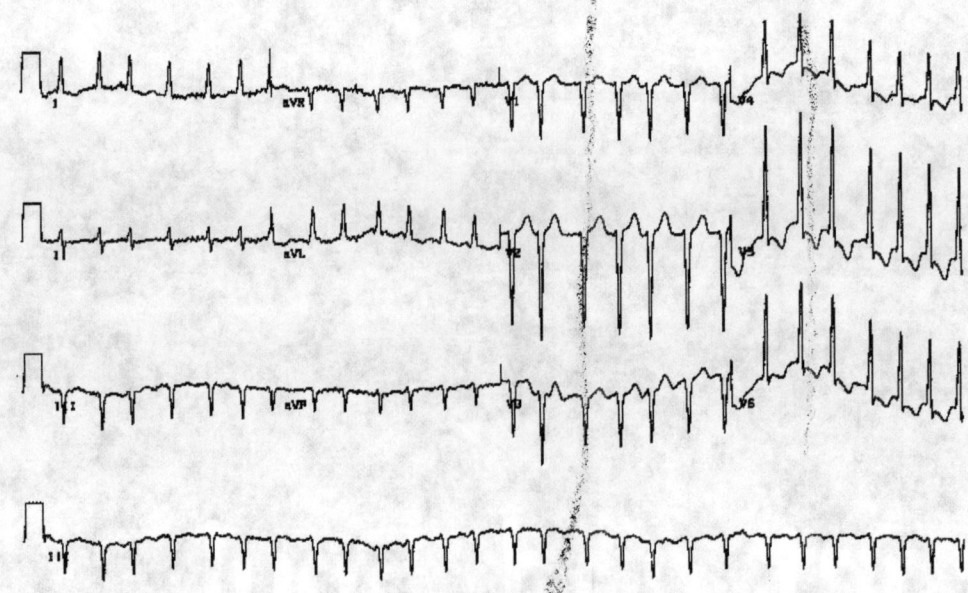

Spend 5 minutes just looking at the ECG above – so often as a doctor these are put in front of you and you are expected to identify the problem and quickly bring to the forefront of your mind all the knowledge you have ever known about it! Try and do this, think about the symptoms the patient might present with, how you might manage it...just let your mind wander....after a few minutes away come back and start working through this section....you will be amazed how much you remembered!

------------------------------------5 minutes------------------------------------

As a general medic you will frequently be expected to manage patients in atrial fibrillation. Once you have done a cardiology job, the treatment of atrial fibrillation will become a walk in the park... until that day you need a simple approach.

There are as I see it three decisions you need to make:

1) What type of atrial fibrillation the patient has?

2) How are you going to treat the atrial fibrillation?

3) How you are going to minimise the risk of thromboembolic complications in these patients?

Let's make decision 1:

To help make this decision, let's first clear up some confusion...

- AF can mean atrial fibrillation, atrial flutter or al fresco – so don't write AF, write what you mean! Lecture over!

- There are specific types of atrial fibrillation and each specific type is defined in a very particular way – it is worth knowing these otherwise you will be left doing the "Ricky nod", you know the nod where your head moves slowly forwards and backwards whilst your mind says "I haven't a clue".

Here is a simplified and well accepted classification:

- First detected episode (irrespective of whether it is symptomatic or self-terminating)
- Recurrent episodes, when a patient has 2 or more episodes of atrial fibrillation
  - If episodes of atrial fibrillation terminate spontaneously then the term **paroxysmal atrial fibrillation** is used. Such episodes last less than 7 days (typically < 24 hours)
  - If the arrhythmia is not self-terminating then the term **persistent atrial fibrillation** is used. Such episodes usually last greater than 7 days
  - In **permanent atrial fibrillation** there is continuous atrial fibrillation which cannot be cardioverted or if attempts to do so are deemed inappropriate. Treatment goals are therefore rate control and anticoagulation if appropriate

Ok, clear as mud? Good!

Decision 2: How are you going to treat the atrial fibrillation?

Ok, now there is no two ways about this, atrial fibrillation treatment is funky! And there is no easy answer and unfortunately consultants differ in their approach... so what do I expect of you... a general understanding and approach! You may remember in the introduction I said there would be areas where I would provide you with some detailed information... well here we go!

The Royal College of Physicians and NICE published guidelines on the management of atrial fibrillation in 2006. The following are also based on the

Joint American Heart Association (AHA), American College of Cardiology (ACC) and European Society of Cardiology (ESC) 2002 guidelines.

Agents with proven efficacy in the pharmacological cardioversion of atrial fibrillation:

- Amiodarone
- Flecainide (if no structural heart disease)
- Others (less commonly used in UK): quinidine, dofetilide, ibutilide, propafenone

Less effective agents:

- Beta-blockers (including sotalol)
- Calcium channel blockers
- Digoxin
- Disopyramide
- Procainamide

Agents used to control rate in patients with atrial fibrillation:

- Beta-blockers
- Calcium channel blockers
- Digoxin (not considered first-line anymore as it is less effective at controlling the heart rate during exercise. However, it is the preferred choice if the patient has coexistent heart failure)

Well that's all well and good but when do we use rate or rhythm control? Good question!

The table on the next page indicates some of the factors which may be considered when deciding on either a rate or rhythm control strategy.

Factors favouring rate control	Factors favouring rhythm control
• Older than 65 years • History of ischaemic heart disease	• Younger than 65 years • Symptomatic • First presentation • Lone atrial fibrillation or atrial fibrillation secondary to a corrected precipitant (e.g. Alcohol) • Congestive heart failure

Right, stop! It has not escaped me that this section is very wordy. Therefore on your return you will be rewarded with a couple of questions that I have adapted from your favourite revision source on examination. Not only should this remind you of happy days gone by but I'm hoping it will reinforce the simple principles that I want you to take away.

Take a break! Watch some TV, I recommend some Gavin and Stacey!

Tidy!

Welcome back to final's revision!

Q.7
A 72-year-old man presents with lethargy and palpitations for the past four or five days. On examination his pulse is 123 bpm irregularly irregular, blood pressure is 118/70 mmHg and his chest is clear. An ECG confirms atrial fibrillation. What is the appropriate drug to control his heart rate?

| A Amiodarone |
| B Atenolol |
| C Digoxin |
| D Amlodipine |
| E Flecainide |

Q.8
A 65-year-old man with a history of paroxysmal atrial fibrillation presents with palpitations. He has no other history of note and a recent echocardiogram was normal. An ECG confirms fast atrial fibrillation. Which one of the following agents is most likely to cardiovert him into sinus rhythm?

A Sotalol
B Procainamide
C Flecainide
D Disopyramide
E Digoxin

A.7

Atenolol (B)

For rate control we know that first line treatment is a beta blocker unless the patient has concomitant heart failure...this patient's chest is clear and therefore heart failure is unlikely. Atenolol is therefore treatment of choice.

A.8

Flecainide (C)

From the information given above we know that the most common drugs used for chemical cardioversion are flecanide and amiodarone... the other things we know about this patient is that they have a normal heart. Therefore flecanide would be indicated.

Principles to take away

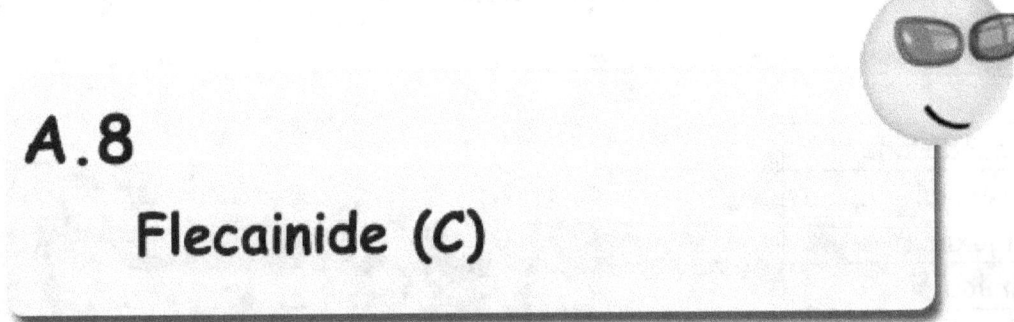

Atrial fibrillation: rate control - beta blockers preferable to digoxin

Atrial fibrillation - cardioversion: amiodarone + flecainide

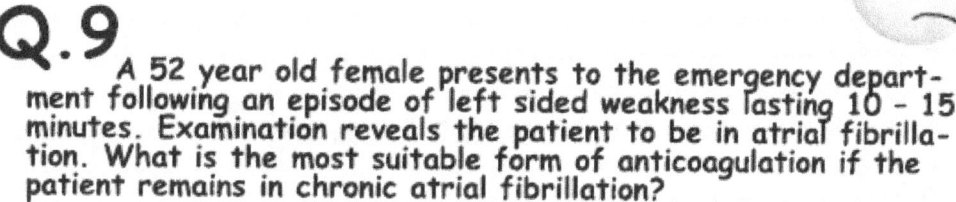

Q.9

A 52 year old female presents to the emergency department following an episode of left sided weakness lasting 10 – 15 minutes. Examination reveals the patient to be in atrial fibrillation. What is the most suitable form of anticoagulation if the patient remains in chronic atrial fibrillation?

There are days in the hospital where it will seem that every patient you see is in atrial fibrillation and unfortunately the decision on whether the patient should be warfarinised or not is a less than scientific one!

However, dismay not my friends, for this decision does not need to be as irregular as the underlying condition... I know, poor joke! So what is this system I hear you cry? CHADS2.

An alternative approach is the **CHADS2** score:

	Condition	Points
C	Congestive heart failure	1
H	Hypertension (or treated hypertension)	1
A	Age > 75 years	1
D	Diabetes	1
S2	Prior Stroke or TIA	2

The table on the next page shows a suggested anticoagulation strategy based on the score.

Score Anticoagulation

0   Aspirin

1   Aspirin or warfarin, depending on patient preference and individual factors

2   Warfarin if not contraindicated

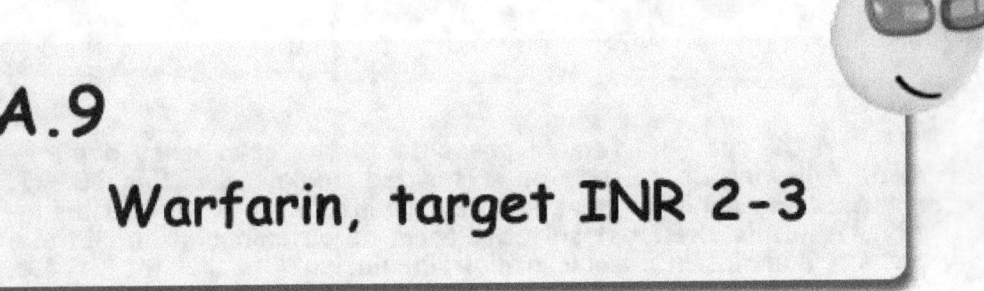

Myocardial infarction (MI)

Look at images below... use them as prompts to bring to your mind all your forgotten knowledge about an MI.

Take 5 minutes to do this.

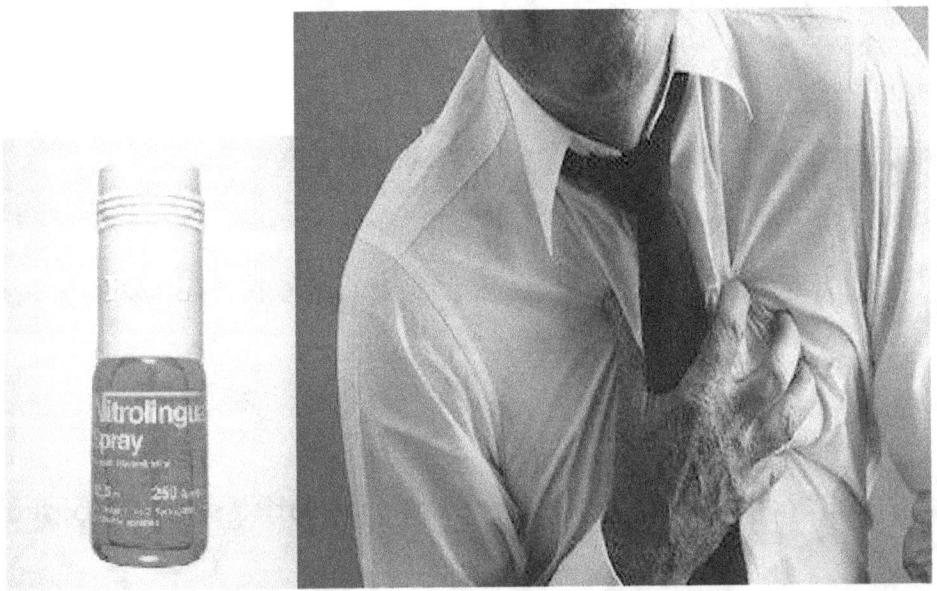

---------------------Take 5 minutes--------------------

58

Now consider this:

> A 57 year old man is admitted with central chest pain radiating down the left side. ECG shows ST depression in the inferior leads and IV morphine is provided with some effect. Previous history is significant for a thrombolysed myocardial infarction 2 years ago, asthma and type 2 diabetes. You provide the patient with oxygen, aspirin, clopidogrel and clexane.

What score should be used to determine if an intravenous glycoprotein IIB/IIIA receptor antagonist should be used?

Ok, so this was a difficult question! I'm sure you are very familiar with the other treatments so I didn't want to bore you with aspirin clopidogrel blah blah blah... if you're not, there will be a brief reminder soon, don't worry! Anyway, you would have heard about it but you probably just didn't bother to learn it... well here's that moment.

TIMI Score (Thrombolysis in myocardial infarction score)

The TIMI risk score is one method used to assess risk in patient with ACS. The TIMI risk score gives a score of 1 for each of the following:

- Age older than 65 years
- More than three coronary artery disease risk factors (hypertension, hyperlipidaemia, family history, diabetes, smoking)
- Known coronary artery disease
- Aspirin use in the last 7 days
- Severe angina (more than 2 episodes of rest pain in 24 hours)
- ST deviation on ECG > 1 mm
- Elevated cardiac markers (CK-MB) or troponin

The risk of myocardial infarction or death within 14 days increases with increased total score:

- Total score 0-2: 3% risk
- Score of 4: 7% risk
- Score of 6-7: 19% risk

High risk patients benefit from and should be given, glycoprotein IIB/IIIA receptor antagonists whether or not they are to undergo a percutaneous coronary intervention.

Now rewrite the TIMI score without looking as I fear again you may have just looked at it and decided it wasn't something that needed learning!!!

Right if that was as unsuccessful as I feared it would be, take a little rest and try again. The next bit is going to be detailed....it's one of those bits of medicine that you need to know on day 1 because you will be asked to treat someone in midst of an MI.

MI management

Management of non ST elevation acute coronary syndrome is based upon the calculation of risk score, for example the TIMI score you now hold dear to your heart.

All patients should receive:

- Aspirin 300mg
- Nitrates or Morphine to relieve chest pain if required

Clopidogrel 300mg and low molecular weight heparin should also be added to higher risk patients. As I am writing this Fondaparinux is being rolled out on the admissions unit, so keep an eye out for this replacing clexane! Once daily – no patient weight required – much easier!

The patient should then be referred to the cardiology team as they may want to stent them.

PCI (Percutaneous coronary intervention) is a technique used to restore myocardial blood flow both in patients with stable angina and acute coronary syndrome.

Following insertion the most important factor to prevent the main complication of stent insertion, notable stent thrombosis is antiplatlet therapy. Aspirin should be continued indefinitely. The length of clopidogrel treatment depends on the type of stent, reason for insertion and consultant preference but is usually 1 year and SHOULD NOT be stopped willy nilly by non cardiology doctors! You've been warned!

Now the basics are revisited, try the challenge below.

Mix and match

For each of the types of MI use the words below to complete the table showing the correlation between ECG changes and coronary artery territories

	ECG changes	Coronary artery territory
Anterioseptal		
Inferior		
Anterolateral		
Lateral		
Posterior		

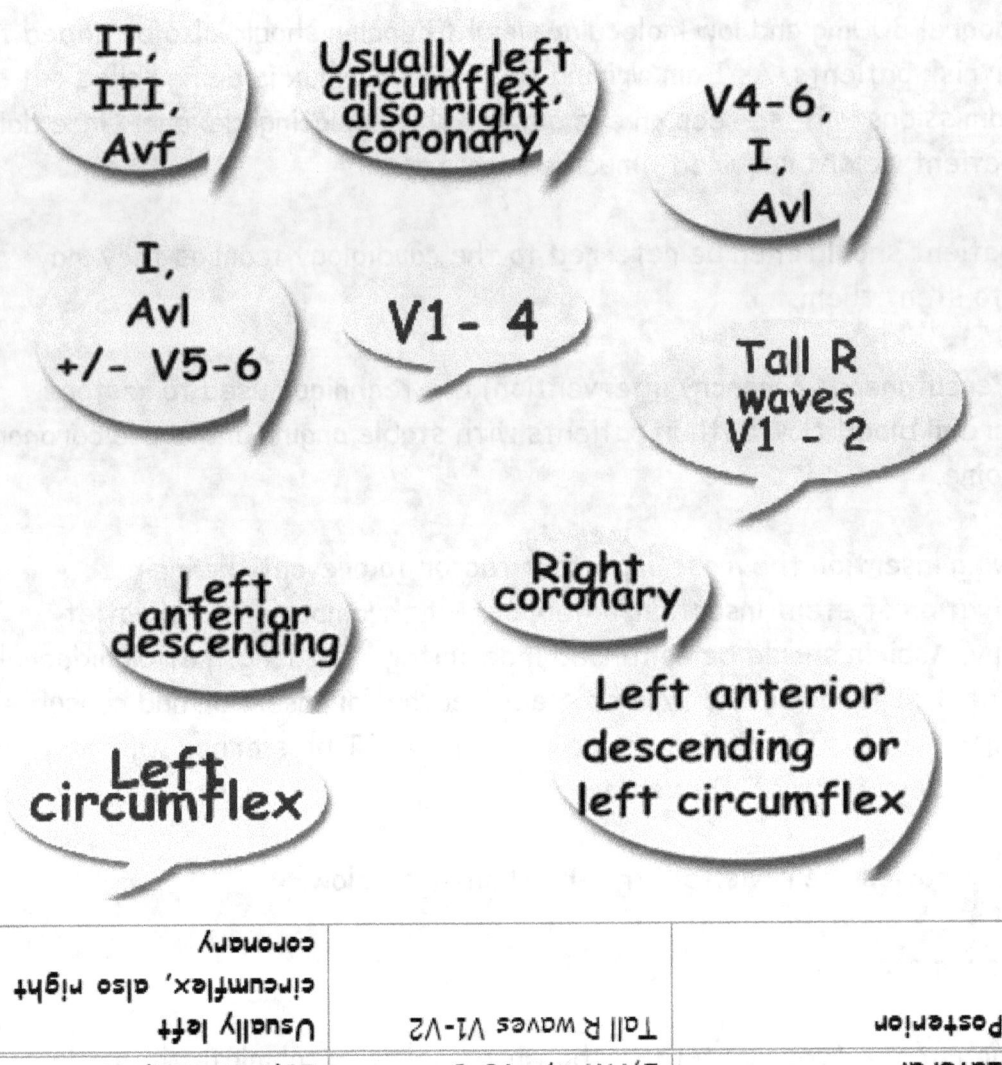

Territory	ECG Changes	Coronary Artery
Anteroseptal	V1 – V4	Left anterior descending
Inferior	II, III, AVf	Right coronary
Anterolateral	V4-6, I, AVl	Left anterior descending or left circumflex
Lateral	I, AVl +/- V5-6	Left Circumflex
Posterior	Tall R waves V1-V2	Usually left circumflex, also right coronary

A word on thrombolysis

You **must** know the indication for thrombolysis – No excuses!

They are:

- ST elevation more than 1mm in two adjacent limb leads
- ST elevation more than 2mm in two adjacent chest leads
- New onset left bundle branch block

PCI has emerged as the gold standard treatment for STEMI but is not available in all centres. Thrombolysis should be performed in patients without access to primary PCI.

With regards to thrombolysis:

- Tissue Plasminogen Activator (tPA) has been shown to offer clear mortality benefits over streptokinase
- Tenectaplase is easier to administer and has been shown to have non-inferior efficacy to alteplase with a similar adverse effect profile

An ECG should be performed 90 minutes following thrombolysis to assess whether there has been a greater than 50% resolution in the ST elevation.

- If there has not been adequate resolution then rescue PCI is superior to repeat thrombolysis
- For patients successfully treated with thrombolysis PCI has been shown to be beneficial but optimal timing remains undecided

Secondary prevention MI

After you have successfully diagnosed and treated your first MI I think you deserve a good night out – and many an alcoholic beverage... I tell you this not only because it's FUN TO DRINK ALCOHOL... though lets be fair it is... but also because it helps me remember what comes next after treatment and that is secondary prevention...makes it sound like the morning after pill! Anyhow, consider the question below.

Q.10

A 61 year old man, normally fit and well, is admitted to hospital with chest pain. He is treated superbly by the FP1, who amazed everyone by reciting the TIMI score. Two months after discharge he is seen routinely in outpatients, which combination of drugs should he be taking?

NICE produced guidelines in 2007 on the management of patients following a myocardial infarction. Some key points are listed below.
All patients should be started on the following drugs:

- ACE inhibitor
- Statin
- Oral nitrate – to be considered
- Beta Blocker
- Aspirin

In case you hadn't noticed, this spells ASOBA (as in not sober... as in drunken post MI treatment)

As regards clopidogrel and aldosterone antagonists:

Clopidogrel

- After an ST segment elevation MI, patients treated with a combination of aspirin and clopidogrel during the first 24 hours should continue this treatment for at least 4 weeks
- After a non ST segment elevation myocardial infarction clopidogrel should be given for the first 12 months

Aldosterone antagonists

- Patients who have had an acute MI and who have symptoms and/or signs of heart failure and left ventricular systolic dysfunction, treatment with an aldosterone antagonist licensed for post MI treatment should be initiated within 3-14 days of the MI, preferably after ACE inhibitor therapy. This my friends is but one reason why most patients should have an ECHO post MI

A.10
ACE inhibitor + Beta Blocker + Statin + Aspirin

End Time!

Well done, you've made it. You're now ready to take the plunge. Good luck on that first month and be safe in the knowledge that we've all been there and we all survived (well, most of us).

Feedback form (please don't stop reading!)

I hope this book provided an enjoyable way to help prepare for those daunting but ultimately exciting first few weeks as a junior doctor. I know that feedback forms are very rarely bothered with, I know how many I threw in the bin as a medical student. Therefore, I have created an online feedback form for you to complete.

There are 3 questions to answer now and another three after you have completed your first couple of months as a junior doctor.

To complete the following feedback forms, visit:

www.rickyfrazer.com

Feedback form 1

What are your overall feelings about the book?

What 3 parts of the book did you most like?

What 3 parts of the book do you think could be improved most?

Further comments

Feedback form 2

How did the book prepare you for your first couple of months as a junior doctor?

Which topics were particularly helpful?

Which other topics based on your experience of your first month would you like to see in the next edition?

Further comments

Notes:

www.ingramcontent.com/pod-product-compliance
Lightning Source LLC
Chambersburg PA
CBHW082350190526

45165CB00022B/2274